Weather and the Seasons

by Margaret McNamara

Table of Contents

Words to Think About 2

Introduction 4

Chapter 1 The Weather in Summer 6

Chapter 2 The Weather in Fall 8

Chapter 3 The Weather in Winter 10

Chapter 4 The Weather in Spring 12

Conclusion 14

Glossary and Index 16

Words to Think About

rain

Umbrellas help us stay dry in the rain.

seasons

Summer, fall, winter, and spring are seasons.

snow

When we see snow, the air is very cold.

temperature

The temperature is how hot or cold something is.

thunderstorm

A thunderstorm has lightning and thunder.

weather

The weather is sunny today.

Introduction

Seasons are times of the year. A year is twelve months long. Each season is about three months long. The seasons are summer, fall, winter, and spring.

▲ What do you like to do in each season?

We can describe each season by talking about the **weather**, what people do, and the hours of daylight.

winter

spring

Chapter 1

The Weather in Summer

The **temperatures** are warmest and the days are longest in summer. Many people like to do things outside.

▲ This family is hiking in summer.

▲ Many people like to swim in summer.

Sometimes **rain** can ruin a summer day. Do you see a lot of rain in summer? Some places have **thunderstorms** that pour down rain.

▲ We hear thunder and see lightning in a thunderstorm.

Chapter 2
The Weather in Fall

Temperatures are cooler in fall, so people often wear pants and shirts with long sleeves.

▲ People play football in fall.

Did You Know?

Some places are very warm in fall.

Miami, Florida ▶

The temperatures are cooler and the days are shorter in this season than they are in summer. What do you like to do in fall?

▲ Some places have hurricanes in early fall.

Chapter 3
The Weather in Winter

Temperatures are coldest in winter, so people dress in warm clothes when they do things outside.

▲ These people are wearing warm clothing in winter.

LOOK AT TEXT STRUCTURE

Cause and Effect

The word "so" helps us understand the effect of cold temperatures. In another sentence, the word "because" tells us a cause. What is the effect?

Because the sun shines for the fewest hours, the days are shortest in this season. Some places have **snow**. What do you like to do when the weather is cold?

▲ A blizzard is a huge snowstorm.

Chapter 4
The Weather in Spring

People say that spring comes in like a lion and goes out like a lamb. Why do you think people say this about spring?

▲ Many people play baseball in spring.

▲ A picnic is a fun way to celebrate spring.

Think about the weather in winter and the weather in summer to help you answer the question. By the end of spring, the temperatures are warmer and the days are longer.

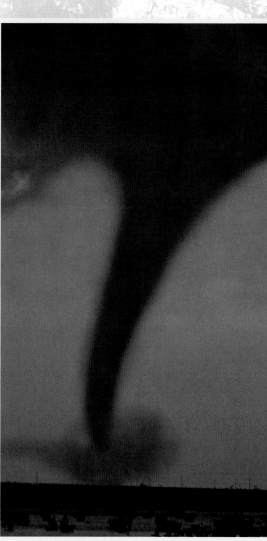

▲ Tornadoes happen mostly in spring.

Conclusion

The weather in each season is different. People wear different clothes and do different things in each season.

summer	fall
• warmest temperatures • longest days	• cooler temperatures • shorter days

winter	spring
• coldest temperatures • shortest days	• warmer temperatures • longer days

Glossary

rain water that falls from clouds

See page 7.

seasons groups of months that have similar weather

See page 4.

snow frozen water that falls from clouds

See page 11.

temperature how hot or cold something is

See page 6.

thunderstorm a storm with lightning and thunder

See page 7.

weather how hot or cold it is; how wet or dry it is

See page 5.

Index

fall, 4, 8–9

rain, 7

seasons, 4–5, 9, 11, 14

snow, 11

spring, 4, 12–13

summer, 4, 6–7, 9, 13

temperature, 6, 8–10, 13

thunderstorm, 7

weather, 5, 11, 13–14

winter, 4, 10–11, 13